Also by Dr. Gloria Gilbère

I was *Poisoned* by my body:

I have a gut feeling you could be too!

The Odyssey of a Doctor Who Reversed Fibromyalgia, Leaky Gut Syndrome and Multiple Chemical Sensitivity, Naturally

(Lucky Press, 2001)

Invisible Illnesses

Non-Drug Solutions to *Understanding* and *Reversing* the Common Denominators in:

- MULTIPLE CHEMICAL SENSITIVITIES
- FIBROMYALGIA
- ALLERGIC/INFLAMMATORY ARTHRITIS
- CHRONIC FATIGUE
- CHEMICALLY INDUCED IMMUNE SYSTEM DISORDERS
- LEAKY GUT/IRRITABLE BOWEL/COLON DISORDERS
- PRESCRIPTION DRUG WITHDRAWAL SYNDROME

(Freedom Press, 2002)

What Consumers Are Saying

"I have been sick with the invisible illnesses of Chronic Fatigue Syndrome, Multiple Chemical Sensitivities, Fibromyalgia, and Bowel Disorders for over thirty years. Consequently, I was intolerant to ANY form of minerals and vitamins. Capra mineral whey was the first form of minerals and electrolytes that my body could tolerate without negative side-effects. I have been using this product regularly now for over one year, and because my body has been made stronger with their use, I have since introduced other supplements to my diet as well. I have noticed a decrease in muscle cramping and an increase in energy and stamina. AND goat whey tastes great too!"

–JENNIFER M., New York, USA

"When I first consulted with Dr. Gilbère for my digestive complaints, it was after several years of getting no relief from the drugs suggested by my medical doctors. I knew there had to be a natural way to help repair my digestion and the damage from all the antibiotics I was prescribed through the years for recurring sinus infections and seasonal allergies. Dr. Gilbère outlined a protocol of detoxification along with a probiotic and mineral replacement from goat's milk and digestive enzymes. Within 3 weeks of starting my protocol, my heartburn was none-existent and my lower G.I. symptoms greatly alleviated. It's now been over 3 years and I'm completely free of symptoms. I have to admit, I was reluctant at first but she made a believer out of me."

–JEFFREY D., E. Sussex, England

"For two years I have been existing on fish and brown rice. Not, of course, because I'm a lover of these two items, but rather, because my chronic yeast infection damaged my gut and that was all my digestive system could tolerate. In time, through the help of Dr. Gilbère, I added goat products such as goat milk, Capra Mineral Whey, and lastly Goatein. It was a wonderful gift to be able to add these foods to my limited diet; I am very grateful. Praise the Lord for giving us the goat!

–WARREN T., Arizona, USA

"Mineral whey is a fabulous and delicious mineral supplementation…a wonderful way to balance electrolytes…it's delicious, I love it"

–CINDY VAN V., New York, USA

"I was having chronic restless leg syndrome when Dr. Gloria told me I needed to start taking Goat Whey. I was desperate for help, so I reluctantly agreed to take it each morning. It took awhile before helping; I did not get immediate complete relief of my symptoms but discovered I was experiencing better overall health so I continued. My Goat Whey has become a morning necessity much like many people have their coffee to "get going."

When my husband developed restless legs, that he sleeps through but keeps me awake, he also started taking the Goat Whey. If he remembers to take a scoop before going to bed, he doesn't keep me awake with his symptoms."

–VERLA MAE S., Idaho

"I am grateful Dr. Gilbère suggested I use Capra Mineral Whey. When she told me to try it to cure muscle spasms that I had in my legs at night, I doubted it would help. After the first day I took the whey I was able to sleep the whole night, without any spasms at all. I have continued to take it and feel a lot better."

–ROSS P., Canada

What Health Professionals Are Saying

"I believe that Capra Mineral Whey is one of the finest acid-neutralizing foods that we can possibly put into our bodies. It is not only high in potassium but is high in sodium and will do a wonderful job in taking care of chronic acid in the stomach wall. Digestion always improves when taking Capra Mineral Whey."

–DR. BERNARD JENSEN, D.C., PH.D., Escondido, CA

"Eating right isn't enough anymore, we need the Capra Mineral whey to supply us with the enzymes and minerals for a healthy digestive track and a healthy lifestyle."

–DR. LAMAR DILTZ, N.D., M.T., Kansas City, MO

"The ages old ability of the goat to provide nutrition suited to man, shines through with Capra Mineral Whey, one of life's greatest foods to promote digestion, synovial activity and sound cell structure."

–DR. PAUL LEATHAM, PH.D., Fillmore, UT

"CMW is helpful in reducing, and many times, eliminating digestive disturbances, painful and inflamed joints, and nerve pain. I personally use the product as well as recommend it to my patients."

–DR. DONALD BODEEN, D.C., Poughkeepsie, N.Y.

"Live-blood microscopy has shown the CMW aids in the reduction of bacterial forms and improves the positive function of blood cell health performance. It neutralizes blood acid levels which otherwise cause various disbiosis to our human system."

–DEANNA BURGESS, CERTIFIED BLOOD ANALYST,
Victoria, BC, Canada

Nature's PRESCRIPTION MILK

Discovering the healing powers of goat-milk products

by Gloria Gilbère, N.D., D.A.Hom., Ph.D.

Freedom Press
Topanga, CA

Bill E. *Goat*

"It's goat to be good...naturally!"

Disclaimer: This book contains nutritional information that, in the author's research and experience, has helped those who have followed it.

The author recognizes that the treatment of illness, and the enhancement of health through detoxification, supplementation and diet should be supervised by a health care professional. Readers should not engage in self-diagnosis and self-treatment. Because each individual case is unique, you are urged to consult your health care professional before embarking on any health regimen.

The author and publisher do not assume medical or legal liability for the use or misuse of the information, products or regimen contained in this book.

The contents of this book are not meant to diagnose, treat or cure any illness, disease or disorder; it is meant for educational and informational purposes.

ISBN 1-893910-23-7
Second printing
Printed in the United States
Published by Freedom Press
1801 Chart Trail
Topanga, CA 90290
Bulk Orders Available: (800) 959-9797
E-mail: info@freedompressonline.com

Dedication

I dedicate this book to the late Dr. Bernard Jensen whose research and work saved the lives of thousands–while furthering a greater understanding and respect for the body's ability to achieve wellness through detoxification, healthy nutrition and his reverence for the magic of goat-milk and goat-milk products.

Table of Contents

The history of medicine: a perspective

2000 BC Here, eat this root and drink this goat's milk

1000 BC That root is heathen; Here, say this prayer.

1805 AD That prayer is superstition; Here, drink this potion.

1940 AD That potion is snake oil; Here, swallow this pill.

1975 AD That pill is ineffective; Here take this antibiotic.

2000 AD That antibiotic doesn't work anymore, it kills all the bacteria–good and bad; Here, eat this root.

2002 AD That root is not being digested; Here take this digestive enzyme and goat's milk product to correct the underlying causes and restore wellness, naturally.

Author unknown
Adapted by Dr. Gilbère

Acknowledgements

AN AUTHOR IS AN ARTIST and an architect. He/she assembles words one-by-one and gradually builds it into a piece of literary art known as a manuscript. Without a supportive, healthy environment, both physical and emotional, the palette of words cannot become an artful finished product–the book. The following people (each in their unique way) have helped create the harmonious space in my life through their encouragement, love, friendship, belief in my work, and in my ability to continue my dream of creating the art I'm so passionate about–teaching the art of wellness through spoken and written work.

Sharon Wiseman–For being my "invisible" builder of support by always anticipating my needs with love and un-selfishness, and by assisting me in making sure my cup is always "half-full" and not "half-empty". Your loyalty, positive consciousness and humor are irreplaceable–I am truly blessed.

Chris Davey–My friend, and fellow author, for strengthening my "wings" by believing in my work and cheering me onward when my "flight" needs a "lift." Your friendship, candor and sensitivity are life-long treasures that are the highlight of this author's journey as I soar through the trade-winds of life, thank you.

Mike Berkich–The only person I know that can talk and think as quick as I do–now we need to work on getting you the same speed on the computer. You've been a loyal friend through my illness and recovery and I'm so blessed for your enthusiastic support of my efforts to share my experiences with all that will listen. I am indebted to you for your patience in educating me on the health benefits of goat-milk products. Your un-wavering dedication to creating quality products and, most of all, noting where the "voids" exist and then creating products to "fill" them is commendable. Working with you has facilitated dreams to mature into reality. It's one thing to be a "dreamer," it's another to be a "dream maker," but it takes a special individual with boundless energy and faith, YOU, to then forge those dreams into action while inspiring others... I can hardly wait to see what's next.

Stan Meyerson–My life is richer because of your generosity in sharing your knowledge of neutraceuticals, your friendship, and belief in my work. You personify the meaning of the word "mentor", thank you for being mine. Your departure has left a huge void in the lives of thousands that you dedicated your life to helping. However, rest assured by dear friend, your work was not in vain, the next generation is proudly continuing your legacy. You are sorely missed, but the gift of yourself is a treasure that will surely grow more valuable in the lives of those you've touched.

Foreword

IF YOU ARE READING THIS, most likely you have a story to tell about your health. Perhaps you are searching for answers to a current health problem or have overcome a condition that bothered you in the past. If the former, Welcome! If the latter, Congratulations! In this book Dr. Gilbère sets forth answers to many health problems and explains how you may overcome them. She's been there. Done that. I think that before you are finished reading this book she and you will seem like old friends.

In the book of Proverbs we are told that "there is nothing new under the sun." Thus the material presented herein is nothing new, that is, new to the world. In fact, it was the great physician from the island of Cos, Hippocrates, who reminded his followers that in all disease it is wise to look first to the spine and the bowel. In our time Dr. J. Bernard Jensen of Escondido, California was a renowned pioneer who took this secret and helped tell it to the world again. You see, there really are no new health secrets under the sun. But if something is new to you, if you have not known about it, experienced it, then it has been effectively hidden from you. It's a secret. One of the "secrets" Dr. Jensen worked so hard to pass on to others was the efficacy of the liberal use of goat milk products in conjunction with any health building regimen and especially his bowel cleansing or "Tissue Cleansing Program," as he called it. It is easy to forget that most of the milk which flowed in the biblical land that was "flowing with milk and honey" was goat milk. It is well to be reminded that most of the world still partakes of the milk of the goat, not the cow, the American Dairy Council notwithstanding!

It was my honor and privilege to study and teach with Dr. J. Bernard Jensen for nearly twenty-five years. As with Dr.

Gilbère, I was drawn to this work in searching for improved health. Genetically speaking, I was not in the front row when the bowels were handed out. Thus, I had to learn how to take care of my bowel in order to regain and maintain my health. You can do the same! Dr. Jensen learned about the bowel from Dr. John Harvey Kellogg of Battle Creek Sanitarium fame, and later from Victor Earl Irons as well as from his study in the anatomy lab of the National College of Chiropractic at Lombard, Illinois. Dr. Gilbère learned many valuable lessons from Dr. Jensen. Although Dr. Jensen is no longer with us, having died in his ninety-second year, I know that he would have been thrilled to see this book!

My loving wife and partner, Joyce Ethel, who has been engaged in bowel work for many years, once commented to Dr. Jensen, "I think people need to bow to the bowel, because the bowel is the king of all the elimination systems of the body." Then she quickly asked, "Wouldn't you bow to a king?" Dr. Jensen immediately liked that metaphor and used it in his books. In this book Dr. Gloria Gilbère lays it all out for you in an easy to understand manner. She knows the bowel is king. She takes seriously Hippocrates' dictum, that in all disease we must look to the bowel. So, my good reader, you have this book in your hand. It is your key to good health. Don't lose it. Get started. The secret's out!

–DONALD VERNER BODEEN, DC, Ph.D., MST

Introduction

THE WRITING OF THIS BOOK CAME ABOUT as a result of hundreds of calls from clients who are victims of what I call "the invisible illnesses" of Fibromyalgia/ME, Chronic Fatigue, Multiple Allergic Response Syndrome (MARS)/ Environmental Illness, Irritable Bowel, Crohns, Intestinal Permeability (leaky gut syndrome), Digestive/Colon Disorders, Prescription Drug Withdrawal Syndrome and Chemically Induced Immune System Disorders. This group of clients is concerned with finding high-quality nutrition that is easily digested and absorbed because the digestive system is by some means compromised.

Victims of invisible illnesses are reacting to many, if not most, food and supplements because the immune system is so severely damaged. When liver function is also compromised, the body is not only starved for good nutrition, it also begins to react to environmental stimuli leading to environmental illness and other autoimmune disorders–this can, and does, change it's victims lives in what seems an instant.

Myself being a victim of leaky gut syndrome and multiple allergic response syndrome (MARS), understand only too well the frustration and fear when everything you eat becomes your enemy. This book will assist the reader in understanding the benefits of organic goat-milk products and their influence on the digestive system–It is for those who are embarking on the road to wellness and those who seek a pro-active approach to maintain health for themselves and their loved ones.

Chapter one

A doctor's personal odyssey from near starvation and anaphylactic shock to wellness

IMAGINE–You're a healthy female in your mid 40's, you maintain a busy professional practice, you travel extensively teaching, lecturing and consulting and you do not experience any physical symptoms of illness or disease. Then suddenly, an accidental fall on concrete in 1993 creates life-threatening blood-clots in your leg that develop into pulmonary embolism within a few days. You spend seven horrifying days in intensive care; the drugs that saved your life almost take your life and leave you with the subsequent after-effects diagnosed as Fibromyalgia, Chronic Fatigue, Constipation and Insomnia. You struggle and keep looking to your physicians for answers, all they can offer is more prescription drugs for muscle pain and spasms, anti-depressants because they're convinced most of your symptoms are psychological and you're just being impatient, and sleeping pills because you're exhausted and the bags under your eyes look like your mascara permanently ran after watching a sad movie–you have, and it's your life! You know that if one more medical doctor tells you "everything is within normal range" someone is going to get hurt! You now realize you're in the same vicious cycle of "symptom-care" and "drug-management" as so many of your clients.

IMAGINE–The drugs that are providing *some* relief from the living hell of invisible pain and inflammation are now bringing on a barrage of side-effects–Tachycardia, Heartburn/Acid Reflux, Acute Insomnia, Chronic Constipation, Tingling and Numbness in your legs and feet and Brain-Fog

(Your mind is writing checks your body can't cash). You're a doctor of natural medicine and you *know* you must find ways to assist your body to repair by slowing getting off the prescription drugs and detoxifying. You also understand that you need to find and use natural non-drug alternatives to deal with the pain, inflammation and digestive damage. You embark on a therapeutic routine like you would outline for your own clients; homeopathics to assist in regulating your heartbeat, digestive enzymes for the heartburn, detoxification therapies for the liver, gallbladder and colon and natural alternatives to pain medication and anti-inflammatory drugs. You finally, after several months, repair some of the damage and you're left with only some soft and connective tissue discomfort that is being managed with natural systemic enzymes. No, you're not 100% but you have quality of life and can resume your normal activities, both personal and professional.

IMAGINE–You're on a business trip and you attend a movie in a nice theatre in 1998. You suddenly, and without warning, experience a sharp, breath-taking pain in your right rib area as if a knife is being driven from your shoulder blade to your breast. Your right hand goes into spasms and you do not have voluntary mobility in your fingers. What is this? Is it a stroke or maybe a pinched nerve? You fly home to consult with your local doctors, chiropractors, physical therapists, and specialists. No one can find a probable cause. Then finally, after nine long months, a neurologist diagnoses thoracic outlet syndrome after extensive, expensive tests. You are in so much pain, you cannot drive your car, comb your hair or even extend your arm enough to get dressed and…the pain is excruciating! You keep echoing to all your doctors and health professionals that it feels as if your rib is piercing through your breast. No one can find any causal effects and nothing shows on x-rays.

You finally agree to take non-steroidal drugs (NSAIDS) for the pain and inflammation. The pain is *never* gone but at least you have *some* quality of life. You also develop acute constipation, which is typical of many pain and anti-inflammatory drugs. Exactly six-weeks later came the final assault as you develop Leaky Gut Syndrome and the resulting Multiple Allergic Response Syndrome (MARS). Your throat is swelling several times a day in anaphylactic reactions, your liver is so swollen it protrudes beyond your right breast and you're losing your hair by the handfuls. Your body is now resisting all food and supplements–over a period of 3 months you loose 52 pounds and you become allergic to everything in your environment.

IMAGINE–You suddenly react to familiar cosmetics you've used for years, other people's perfumes or anything fragranced, the exhaust of cars on the road, new paving, refueling your car, the newer furnishing and rugs in your home and office, even the simple act of entering a store to buy groceries or being in the same office as a new computer or copier, all trigger the same life-threatening, terrifying response. Just as serious are the unsuspecting odors in your vicinity because, although you're not sure what it is–it doesn't matter, it still causes a life-threatening allergic reaction. You *know* anaphylactic shock can kill; will the next allergic attack be your last?

IMAGINE–Your entire life and career is staring you in the face. You remember all the clients you've consulted with and the horror stories of how their lives were shattered. You search for healthy, natural products and locate several nutritional products that facilitate reduction of your symptoms and eventually your recovery. People that know you, and watched your decline, now see your dramatic improvement and ask what you did to conquer the invisible illnesses that nearly took your life.

This is *not* science fiction. It is a life, my life. This is my odyssey from pain and near starvation to recovery. It has been a long, arduous road, but a journey well worth it for many reasons.

Just as one specific incident did not cause my invisible illness, one specific product or therapy didn't provide all the answers. However, I credit one group of products for speeding my recovery, and doing so *without the allergic reactions* experienced by previously consumed nutritional supplementation.

Part two

The magic of goat's milk

My healing, with a lot of help from my friends, four-legged ones that is

IN 1998, AT THE PEAK of my second round of illness, I resorted to a diet of freshly juiced organic carrots, apples, ginger, broccoli and brown rice; my body could not tolerate any protein because of the liver damage. I still had the excruciating pain and swelling at the base of my neck that was the size of a large grapefruit. What I eventually discovered was literally a miracle. A client of mine referred me to a therapist, Beata Golau, who utilizes a unique form of lymphatic massage and body-work. I resisted trying one more therapy, but eventually agreed out of sheer desperation.

As soon as I sat on her therapy table and described my pain and symptoms she commented. "Oh! It sounds like your rib is dislocated." I couldn't believe my ears; someone *finally* understood what I had been echoing for months. She proceeded to use a very gentle form of soft tissue therapy with specific applied pressure and we both heard my rib go back into position…the pain of so many months immediately disappeared as described in detail in my book *I was Poisoned by my body* (Lucky Press, 2001). It *had* been a displaced rib all along and now I had to deal with re-building my body, *again*, from the devastating effects of the drugs for pain and inflammation. She then used a unique form of lymphatic drainage massage to eliminate the acute swelling at the base of my neck where fluids had built. The results were amazing! I must admit–I was angry, after so many years of assisting clients to wellness, I was unable to help myself because no one found the actual cause of my pain and no one *really listened* to my description of the pain. This should be a lesson to all of us, especially health-care professionals, to listen, listen, listen, when a patient describes their discomfort; don't form opinions just because the symptoms don't fit the "conventional" model. A little extra effort and investigation could have saved me from the nightmare of prescription drug side-effects that almost took my life un-necessarily. It took months of therapy to get my rib to stay in place longer than a week because of the lost muscle tone and tissue damage. However, I no longer have problems with it dislocating unless I lift something heavy.

Now that my rib was repositioned, I had to find a way to re-build my health and strengthen the connective tissue so the rib would stay in place. Because of my severely restricted diet and the colon detoxification necessary to repair the leaky gut and damaged liver, I was extremely deficient in all nutrients, especially minerals, electrolytes and protein. My colon therapist introduced me to a magic brown powder of a mineral and electrolyte replacement made from organic goat's milk called Capra Mineral Whey™. I felt the immediate effects of renewed energy, calmness in the gastrointestinal tract and reduction of muscle spasms. I then remembered reading books by the late Dr. Bernard Jensen (world renown nutritionist who I had the honor of meeting in 1986) and how milk supplied by a goat (his goat's name was STAR) literally saved his life. I then located a local source for fresh goat-milk, and other goat products, and was pleasantly surprised that my body could not only tolerate it, but also assimilated the nutrients with ease. The immediate health benefits were remarkable, Dr. Jensen was right! I soon started to regain my weight and my muscle tone improved, as did my stamina. I had no allergic responses to goat-milk products nor did they cause any gastrointestinal distress.

Products made from goat's milk are easily tolerated because goat's milk is most like human milk in composition and is praised because it's readily digestible. Our bodies can digest goat's milk in just 20 minutes; conversely, it takes two to three hours to digest cow's milk. Goat's milk is nutritionally balanced to nourish a kid (baby goat) weight of 6–9 lbs., about the same as that of a human baby. The high nutritional quality and the greater ease of digestion of goat-milk is what make it so effective in the prevention or reversal (as in my case) of disease.

The secret lies in its dual natural ability to be *half-cleansing* and *half-building*. Currently, I use products made from goat's milk produced in farms owned by a family business in Chehalis, Washington –They're not new at this, they've specialized in quality organic goat-milk products since 1928. Their goats are never fed any pesticides, herbicides, antibiotics or growth hormones, thus insuring the integrity of this magnificent animal's milk.

Goat's milk contains a unique essential balance of vitamins, minerals, proteins, carbohydrates, enzymes and fats–it is one of the few *whole foods* that are actually *able to sustain life*. Some doctors (and I'm one of them) call goat's milk and goat-milk products "Prescription Milk." It's gained this reputation because doctors and nutritionists worldwide prescribe it for everything from infant colic to advanced degenerative diseases.

A Prize Fight
That's NO Match

Dr. William Lee Secor, a medical doctor from Kerrville, Texas wrote, "I am a surgeon, restricting my practice to surgery, but since discovering the value of goat-milk in the treatment of stomach and intestinal ulcers, I have not operated on a single ulcer case, and this covers a period of a dozen years.

Goat's milk vs cow's milk

I am constantly being asked why I favor goat's milk over cow's milk? Here's a historical perspective that will answer a multitude of questions:

- Nutrient content of goat milk is slightly less than cow's milk, however, more digestible because the fat molecules are one-fifth the size of those from cow's milk–making it easily tolerated by those with compromised digestive systems
- 72% of the milk used throughout the world is from goats; it is one-third richer than cow's milk, yet more nourishing and easier to digest
- The flavor of goat's milk is comparable to that of cow's milk
- Goat-milk has no cream separation because of smaller fat molecules
- Goat-milk contains *pre*-formed Vitamin A that allows it to be readily available for use by the body; it is contained in the milk fat.
- Goat's milk contains a more highly-evolved cholesterol than cow's milk, making it more available for absorption to the brain and body; cholesterol is essential to the health of the myelin sheaths "white matter" of the nerves in the brain
- Goat's milk is closer to human milk and is therefore easily accepted especially by those young or frail
- Goat's milk has an alkaline reaction, the same as mother's milk; cow's milk has an acid reaction

- Goat-milk does not form mucous (phlegm), therefore better tolerated by asthmatics and those with allergies
- Goat-milk contains more chlorine, fluorine and silicon than any other domestic livestock; chlorine and fluorine are natural germicides and fluorine assists in preventing diabetes
- Goat-milk contains 2% curd, which precipitates in the stomach; cow's milk is 10% curd
- Goat's are naturally immune to diseases, such as tuberculosis, and are used in third-world countries to actually cure tuberculosis because of their inherent antibodies
- Goat-milk is tolerated by a compromised/damaged liver because of the smaller fat molecules and…it's naturally homogenized
- Goat-milk has the ability to "sweeten" the intestinal tract and assist with constipation
- Goat-milk contains a higher evolved carotene (pro-Vitamin A)–researchers have found this to have cancer preventing properties

Note: Yes, I must admit, I do have a new love affair–his name is Billy, Bill E. Goat, that is!

Bill E. *Goat*

Part three

Nature's "treasure mine" of minerals

IN THIS AGE OF READILY available global information, it is confusing (to say the least) to both patient and doctor when they read articles and claims about mineral and electrolyte replacements. In my personal and clinical experience, I observed supplementation in many forms and compositions and there is tremendous confusion regarding nutrition. Many argue that if you eat right you shouldn't need supplements. This attitude exists because of confusion between diet and nutrition–*Our diet is the total sum of what we consume; nutrition is the benefits of what we consume.*

A whey to health, naturally

It is important to keep in mind that it's not only what we eat but how our body utilizes what we consume and what is excreted unused.

It is only recently that the role of minerals has received the importance it deserves. The following are examples of disorders, but not limited to, those that are either caused by or affected by mineral deficiencies and imbalances:

- Cardiovascular
- Gastrointestinal
- Muscular
- Skeletal
- Neurological
- Immune
- Endocrine

Minerals do not work alone; they must work in synchronicity (in balance) with one another, as well as with the metabolism of

FYI...
It takes 23 ounces of fresh goat-milk to get the nutritional equivalent contained in 1 oz. of this goat whey powder.

proteins, carbohydrates, fats and vitamins. When one mineral is excessive, it steals from others, thereby causing a chain-reaction of deficiencies.

Examples of minerals "out of balance"

- Excessive Calcium = Loss of Magnesium & Zinc
- Excessive Sodium & Potassium = Deficiency of Calcium & Magnesium
- Excessive Calcium & Magnesium = Deficiency of Sodium & Potassium
- Excessive Sodium = Loss of Potassium
- Excessive Potassium = Loss of Sodium
- Excessive Copper = Loss of Zinc
- Excessive Zinc = Loss of Copper & Iron
- Excessive Phosphorus = Loss of Calcium

I personally experienced the effects of minerals that were "out of balance". During my illness, as in most chronic disorders, a condition of mal-absorption exists. This means the body may be taking in the needed supplement but is unable to make use of it. Therefore, it is either excreted and provides no benefit, or it is stored and can have negative effects with compounding symptoms that are challenging to diagnose. Whenever I consumed minerals or calcium replacements, my right eye would "flutter" or "twitch" to the point where it caused me great concern–not to

"You need to know..." Vitamin "C" can form chelates in the bowel with known metals such as zinc, copper, iron and magnesium. This literally means the chelated metals are prevented from being assimilated through the bowel wall and eliminated.

Conclusion: Do NOT Take Vitamin "C" at the same time you take your mineral supplement

mention the annoyance. I had developed mal-absorption from a damaged intestinal tract caused by a condition known as leaky gut syndrome. My body was not able to utilize the supplements and was storing them in un-safe amounts.

In my quest for wellness, I experimented with over 11 mineral replacement supplements, all without sustained relief. I've also consulted with thousands of clients, world-wide, whose symptoms range from brain-fog (inability to concentrate or connect thoughts) to muscle cramps and involuntary movements. None of their mineral/electrolyte supplements provided complete relief, if any.

After I was introduced to a mineral/electrolyte supplement made from organic goat's milk, much to my delight, I was symptom-free–as were my clients with many of the same complaints.

Your body's electrical system

Minerals are the body's electrical transmitters–make sure you're body is sending the right messages! Minerals serve as electrical signals to every cell, and brain signals are transmitted through body fluids. Intestinal disorders, particularly leaky gut and diarrhea, cause deficiencies of minerals. Since minerals are stored primarily in the body's bone and muscle tissue, the absorbed mineral must be carried by the blood to the cells and then absorbed by the cell membrane to be utilized.

Individuals with inflammatory and circulatory disorders like Fibromyalgia, Peripheral Neuropathy, Diabetes and Hypothyroidism especially experience muscle spasms, tingling, numbness and "brain-fog." These symptoms can be caused, or accelerated, by a deficiency of minerals and the compromised ability for absorption.

The supplement I found superior in "re-connecting" the body's electrical system is Capra Mineral Whey™. This powder is a mineral-rich, golden brown, dry natural food dehydrated from organic goat-milk whey. The nutritional analysis of whey confirms it to be high in minerals, trace minerals and contains some vitamins.

The role of minerals in detoxification

When an individual is undergoing any type of detoxification program, it is essential to replenish mineral reserves. Cleansing/detoxification therapies are life-saving and health-building, however, in the process not only eliminate toxins but also minerals and electrolytes (as in the life-enhancing detoxification gained through colon hydro-therapy). It is just as important to remember that we also loose minerals through exercise, sweating, and consumption of laxatives. Minerals/electrolytes are also acutely depleted during such medical procedures as kidney-dialysis. These electrolytes make-up the electrically charged ions that help regulate water balance, acid-alkaline balance, osmotic pressure, nerve impulse conduction, and muscle contraction.

Capra Mineral Whey™ contains the mineral combinations found in whole foods; high in sodium, potassium and calcium and has an alkaline reaction. The powder is grainy but naturally sweet and

"You need to know…"

Anyone with a chronic health condition, soft and connective tissue pain, muscle cramps ("Charlie-horses"), excessive stress, digestive disorders, lack of mental alertness or depletion of minerals from colon or kidney disorders will benefit from Capra Mineral Whey™. For acute mineral depletion, supplement with 2 Tbs. twice a day. For maintenance use 2 Tbs. once a day. For those in colon hydro-therapy, you should also consider supplementing with mineral whey via rectal infusion. Refer to Resource Section for product availability.

tasty. To me, the taste is like a combination of Ovaltine™ and Graham Crackers™. It contains a jam-packed array of minerals and electrolytes in each serving. It is easily mixed in cool or warm water and makes a delicious energy drink–I prefer it in warm water to be consumed like a broth or tea.

Many of my clients sprinkle it on hot cereal as a natural sweetener instead of brown sugar. It is especially health promoting for the very young and the frail. Mineral Whey is not only essential in restoring the body's electrical balance, but also in maintaining it, Naturally.

Minerals and fiber supplements

Everywhere we look, from magazines and newspapers to radio and television, fiber and fiber products are widely talked about and highly advertised. We are literally being bombarded with commercials and advertisements telling us that eating more fiber is not only the "right" thing to do, but will solve practically all of life's problems. Yes, fiber is essential, and without it any intestinal detoxification therapy cannot be successful. However, it is important to know that fiber supplements should *not* be taken at the same time as mineral/electrolyte replacements–fiber can pull-out the minerals you've just replaced. Therefore, if you take a fiber supplement in the evening, take your mineral replacement early in the day or during the day as a natural "pick-me-up."

Additional benefits of mineral whey

- **ACID NEUTRALIZER**–Takes care of chronic acid in the stomach

- **REPLENISHES DEPLETED RESERVES**– Most valuable in replenishing reserves that are nutritionally depleted through every day stresses and poor food choices
- **ESTABLISHES HOMEOSTASIS—** Minerals are necessary for all systemic functions
- **SUPPORTS WEAK & PAINFUL JOINTS**–Provides the minerals necessary for healthy bones.
- **SUPPORTS ATHLETES**–By replacing minerals lost due to sweating & exertion thus preventing muscle spasms and weakness while promoting endurance
- **SUPPORTS INTESTINAL TRACT**–By assisting to create an alkaline environment that is not friendly to bacteria and viruses
- **SUPPLIES MORE THAN 20 NATURALLY OCCURRING MINERALS**–Is a complex whole food
- **100% BIO-AVAILABLE**–Since whey is pre-digested, it is not dependent on the body's ability to break it down, therefore it is readily absorbed

Composition of goat's milk mineral whey

2-Tablespoons of Capra Mineral Whey™ contains:

• 884 mg Potassium	• 189 mg Sodium	• 313 mg Calcium
• 280 mg Chloride	• 78 mg Zinc	• 249 mg Phosphorus
• 24 mg Magnesium	• 58 mcg Molybdenum	• 1.5 mcg Manganese
• 63 mcg Tin	• 27 mcg Strontium	• 17 mcg Copper
• 12 mcg Chromium	• 8 mcg Antimony	• 8 mcg Nickel
• 8 mcg Silicon	• 2.2 mcg Selenium	• .51 mg Iron
• 52 mcg Barium	• 10 mcg Silver	• 2.8 mcg Vanadium
• 9.4 mcg Titanium	• 1.0 mcg Yttrium	• 17 mcg Sulfur
• 104 mcg Boron		

Minerals and their functions

- **Calcium**–Builds bones and teeth, muscle contraction, heart action, nerve impulses, blood clotting
- **Magnesium**–Bones, liver, muscles, transfer of intercellular water, alkaline balance, neuromuscular activity
- **Sodium**–Electrolyte balance, body fluid volume, nerve impulse condition
- **Potassium**–Cell membrane potential, nerve impulse conduction, heart rhythm, acid base balance
- **Copper**–Hemoglobin formation, production of RNA, elastic tissue formation, cholesterol utilization, oxidase enzyme activator
- **Zinc**–Protein synthesis, carbon dioxide transport, sexual function, insulin storage, carbohydrate metabolism, wound healing
- **Iron**–Hemoglobin formation, electron transport, oxygen transport, enzyme activator
- **Manganese**–Carbohydrate metabolism, protein metabolism, connective tissue, joint fluid production, nerve tissue, Vitamin B_1 utilization
- **Chromium**–Glucose utilization, insulin activity, heart muscle, cholesterol utilization
- **Phosphorus**–Builds bones and teeth, energy production, acid-base balance, necessary in metabolism and cell membranes, calcium absorption

"Did you know?..."

Minerals are called the "Spark Plugs of Life" according to Dr. David L. Watts, because "they are required to activate hundreds of enzyme reactions within the body. Minerals are more important than vitamins since plants manufacture vitamins, but minerals must be obtained from the soil," or from a source such as mineral whey. Life is dependent upon the body's ability to utilize these minerals.

- **Molybdenum**–Enzyme activator, uric acid formation, oxidative enzymes
- **Selenium**–Peroxidase scavenger, glutathione peroxidase, anticarcinogenic, Vitamin E synergist
- **Silicon**–Bone formation, collagen formation, cartilage formation, elastic tissue
- **Boron**–reduces calcium loss
- **Vanadium**–Strength of bones and teeth, lowers blood lipids, inhibits cholesterol synthesis
- **Tin**–Growth, protein synthesis
- **Nickel**–Activates some liver enzymes

Part four

Protein is protein, right?

Life without protein is NOT possible

APART FROM WATER, the most profuse substance in the body is contained in the amino acid group; the matrix into which these substances are incorporated is protein. Growth, development and function depend on the correct supply of amino acids. The brain is made-up of more than one-third protein. Researchers have been able to show that manifestations of symptoms such as depression, apathy, irritability, hyperactivity etc. can occur when non-essential amino acids cannot be adequately produced to meet the demands of the brain.

Protein malnutrition is the frequent precursor of amino acid deficiency. These deficiency-states are often times associated with inadequate diet, malabsorption, compromised digestion, stressful situations, infection, trauma, drug use, imbalances of other nutrients and minerals, and in general, aging.

Proteins are the body's building blocks and the body utilizes approximately 50,000 different proteins to form organs, nerves, muscles and skin. A specialized form of proteins–enzymes–are managers and catalysts in all our biochemical processes; antibodies are also a form of specialized proteins.

Proteins are simply combinations of 22 amino acids, nine of which are "essential" for humans. A nutrient labeled "essential" means it *cannot* be made by the human body and must be supplemented.

When the "essential" amino acids are present in the body, the healthy body can build the "non-essential" amino acids. However, that said, if even one "essential" amino acid is missing or extremely low in concentration, the body is unable to synthesize the other proteins it needs–even in cases where overall protein intake is high.

Protein is "essential" for...

- Normal growth
- Blood Clotting Process
- Formation of milk during lactation
- Regulation of acid-alkaline balance in tissues and blood
- Fighting illness by making red blood cells
- Regulating bodily functions by producing hormones
- Transporting oxygen
- Growth and repair of tissues
- Formation and maintenance of blood, muscle, skin and bone
- Facilitating the transport of iron and copper throughout the body
- Strong, healthy hair and nails

Signs of protein deficiency

- Grayish complexion
- Hair Loss
- Fragile, splitting and slow-growing fingernails
- Unusual fatigue
- Frequent Illness due to compromised immunity
- Females may stop menstruating

FYI...
The only complete source of protein is derived from animal protein; vegetable sources, such as soy and legumes, contain only incomplete proteins.

Nature's most digestible protein

At the onset of my illness and subsequent leaky gut and multiple allergic response syndrome (MARS), there was no way I could tolerate animal protein–the liver was far too damaged to process the molecules.

After I began to recover, and my liver and intestinal ecology improved, I needed a source of pure complete protein. I was losing most of my muscle tone without protein, even though my overall health greatly improved. I also needed protein to maintain muscle integrity so my ribs would stay in place. The day finally came when I was willing to attempt small amounts of goat's milk protein. I was apprehensive, as are most victims of allergic reactions, to consume too much too soon and end up with yet another anaphylactic reaction. I started slowly with Goatein™, a product I recommend to my clients with invisible illnesses, because of its high rate of tolerance. I used one-half teaspoon in rice milk, once a day. Since no allergic reaction was experienced, I soon increased the amount to one teaspoon and then two tablespoons. Goatein™ became the basis for all my liquid meal supplementation, usually with the addition of the Capra Mineral Whey™. Within thirty days my muscle tone improved, as did my energy level and strength. My therapist could feel the difference during my lymphatic massage and my rib was not displacing as often.

FYI...
Goatein™ is fermented with lactic acid which is essential for proper pH balance of the gastrointestinal tract and tissue.

Why did I choose Goatein™ over other protein supplements? Because it is Pure. It is the highest quality protein powder available. It is produced from goat-milk that contains NO antibiotics or female growth hormones. Most commercially available dairy proteins are produced from animals given antibiotics and/or BrST that can transfer to the user unacceptable amounts of growth hormones. Many vegetable protein powders, such as soy, are made from non-organic or GMO sources and may contain residues of chemicals.

Why did I choose animal protein from goat's rather than soy?

Because animal protein is the only complete source of protein available. Animal sources of protein, such as Goatein™ have many advantages over vegetarian sources because vegetarian sources are typically low in one or more of the essential amino acids–even when overall protein content is high.

Today there exists controversy regarding the safety of soy. Soy protein is thought by some researchers to be high in mineral-blocking phytates and thyroid-depressing phyto-estrogens. Soy also contains potent enzyme inhibitors that may even depress growth. It is my belief, and that of some researchers, that ethnic gene make-up should be taken into consideration more than we do. Individuals with Asian decent appear to have the "gene pool" that does well with quality organic soy products. However, as north Americans, our genes do not have the same ability to utilize soy products. For example: Let's say a person of Asian heritage is suddenly in a position where they

FYI...

According to Dr. Patrick Quillin, "The immune system is literally a scoreboard of a person's nutrient intake, since dietary nutrients provide the raw materials (precursors) to build and maintain the immune system. The bulk of immune bodies are composed of protein, which explains why a protein deficiency often leads to depressed immune function. Many cancer warriors, including interleukin and interferon are all measurably reduced in protein malnutrition."

must consume South American food of black beans and garbanzos. Those individuals will usually develop digestive disorders, because their genetic make-up does not have the inherent ability to break-down those proteins–the same is true vice-versa. As a doctor who consults with people around-the-world, I always take into consideration the ethnic gene make-up of my client before recommending dietary life-style changes. This research has been highlighted by work from respected doctors such as Dr. Peter J. D'Adamo. In his book *Eat Right 4 Your Type* he explains, "A chemical reaction occurs between your blood and the foods you eat. This reaction is part of your genetic inheritance. It is amazing but true that today, in the late twentieth century, your immune and digestive systems still maintain a favoritism for foods that your blood type ancestors ate." Interestingly, according to Dr. D'Adamo's book, the Chinese, Japanese, and many other Asian societies are primarily composed of blood Type B. According to this methodology, people with Type B blood greatly benefit from products made of goat's milk.

"You need to know..."

Many scientists claim that high-protein diets cause bone loss. These claims are neither supported by scientific research nor by anthropological surveys, according to Dr. Emmanuel Cheraskin in corroboration with Dr. Weston Price's observations.

According to Dr. Cheraskin, "Inadequate protein intake leads to loss of myocardial muscle and may, therefore, contribute to coronary heart disease. However, protein cannot be adequately utilized without dietary fats. That is why protein and fats occur together in eggs, milk, fish and meats. A high protein, low-fat diet can cause many problems including too rapid growth and depletion of vitamin A and D reserves."

How is goat protein absorbed verses cow protein?

Goat's milk contains smaller molecules that are closer in size and composition to human milk, making it naturally easier to digest. This is extremely important especially with damaged liver function as in my case. Eating other types of protein produced immediate liver pain, however, consuming goat-milk protein did not. Goat-milk protein is also substantially less allergenic than cow's milk protein in highly allergic individuals.

Additionally, Goatein™ is partially pre-digested through a lactic acid fermentation process to make it even more bioavailable while virtually eliminating its lactose content. Research shows that proteins from milk that has been predigested by lacto-fermentation are absorbed more efficiently as compared to those of non-fermented milk. During fermentation, many digestive enzymes are created that aid in the assimilation of food nutrients. According to William Campbell Douglass, M.D., protein contained in cultured (lacto-fermented) dairy products is the very highest quality available for human consumption.

What makes the processing of Goatein™ unique?

Most milk protein powders are made from cow's milk, and even though they claim to be "minimally processed," they use several invasive processing steps that may include heating at high temperatures. These methods denature many important amino acids and destroy enzymes and beneficial bacteria. Research suggests that processing whey with heat and acid (i.e. ion-exchanged) results in the loss of several key amino acids including Cysteine, Threonine, Serine, and Lysine. Because

Goatein™ is processed without the use of acid or excessive heat, the amino acids, enzymes and beneficial bacteria remain in their natural form.

Why I prefer Goatein™ as a protein supplement

Goatein™ is:

- The only protein powder that contains beneficial bacteria designed to be resistant to heat, cold, chlorine, fluorine, stomach acid and extremes of pH.
- A balanced combination of whey protein and milk protein. Research on milk proteins indicates that consuming whey protein by it self can cause too rapid an absorption of amino acids. These amino acids are then used for energy production instead of tissue building. I am convinced that a combination of both whey and milk protein are superior.
- Made by fermentation with lactic acid bacteria that creates biologically active lactic acid that plays a major role in energy production and fat burning.
- Lacto-fermented, a process credited by the healthiest and longest living people for their longevity.
- A cultured (lacto-fermented) dairy product, a process credited by some as capable of lowering cholesterol and protecting against bone loss.

"Do you know?..."

According to research by nutritionist, Robert Crayhon, protein turns into protective enzymes in your body known as CP-450. Yes, it sounds like a World War II airplane, however, "it's the name of an enzyme your body uses to neutralize toxins." In a state of protein deficiency, the body does not produce enough CP-450, therefore, the ability for this substance to eliminate many toxins, and free-radical-producing substances, does not exist. "Low Levels of CP-450 have been associated with increased breast cancer rates. CP-450 is just one of the many guardian enzymes that are made from protein."

- Abundant in natural occurring digestive enzymes including protease, amylase, lipase and lactase.
- Often tolerated by people who cannot drink cow's milk
- Produced from goats that are not fed with pesticide-contaminated food, herbicides, growth hormones or antibiotics.

Part five

Colostrum is *not* a Greek theatre; although it does "set the stage" for the immune system

Immunological cocktail: nature's gift

MUCH HAS BEEN WRITTEN regarding the health benefits of colostrum. I hope to shed some light on the subject by not only discussing its benefits but also why, in my opinion and experience, goat's milk colostrum is superior to cow's milk colostrums in supporting the immune system.

All disease and illness stem, in one way or another, from a weakened immune system that loses its ability to protect. The immune system not only protects us from invasive substances such as yeasts, bacteria and viruses but also from toxic substances ingested as well as environmental exposures. It is the sum of those intelligent, highly trained biological defenders that are inherently trained to identify, attack and destroy disease-causing enemies. It also has a built-in cellular memory that rarely forgets an enemy, even if somewhat mis-shapen or disguised to evade detection. Let's say you are immunized against a specific disease like tuberculosis. The introduction of the tuberculosis germ into the body is not recognized by the immune system; therefore, its surveillance system identifies it as an invader (antigen). Antigens are foreign substances that trigger the body's army with a call to "war," with specific orders to "destroy" by producing substances called antibodies–meaning against foreign or unrecognizable bodies. These antibodies are specific only to that antigen.

If the immune system is destroyed or damaged by, lets say,

drugs, excessive stress, environmental toxins or chemotherapy, it looses its ability to fully protect us against invading enemies. Likewise, if our army gets paranoid and over-zealous during their surveillance maneuvers, they can target healthy cells for destruction–thus the birth of auto-immune diseases like rheumatoid arthritis, lupus, Fibromyalgia and scleroderma, to mention a few.

There is also another form of destruction by our immune cell "army" besides total destruction–it's called phagocytosis, a process by which the invader is eaten or digested, a kind of metabolic cannibalism. This occurs when the body's surveillance cells, which are called phagocytes and include granulocytes and macrophages, signal the immune system response team to destroy as many invaders as possible–viruses, bacteria and dead cells. This type of cell does not have receptors like other immune cells, therefore, communication is somewhat subject to interpretation and invaders are "disposed" of but not completely annihilated. These cells do not need to directly communicate to get their orders, they are natural killer cells (NK cells) and are "programmed" to make decisions on the "spot," on their own; they do not contain a "memory" like antibody cells. They have to eliminate the intruder by releasing toxic enzymes that destroy. These NK cells also have the unique ability to stop viruses from replicating by releasing a substance called interferon, a kind of "metabolic nerve gas."

FYI...
Only natural immunity is transferable; immunity acquired through vaccination is not. Therefore, just because a mother is vaccinated, it does not mean that protection is passed to her child.

The highly sophisticated immune system contains various weapons of mass destruction that, when used against an invading enemy of disease cells, can save a life–when used against its own population of healthy cells, destroys life.

The immune system does not operate solely on its own; it is assisted by the lymphatic and circulatory systems, which are intricate roadways that circulate throughout the body.

The good news–I'm going to share with you ways to send in reinforcement troops through nature's own immunological cocktail, colostrum.

Exactly what is colostrum?

It is the fluid manufactured by the body just prior to the production of mother's milk during the first 24–48 hours after birth.

Research confirms Colostrum has powerful, life-supporting immune growth factors that ensure the health and vitality of the newborn. Colostrum assists the body into "homeostasis," a natural state of equilibrium. It is considered the "perfect food" and is produced by all female mammals just before they give birth. The fluid flows into the mother's milk for about three days following birth; after three days colostrum flow ceases.

How does colostrum work?

Medical research has confirmed that colostrums provides the host (you) with numerous immune factors including a wide spectrum of immunoglobulins , anti-bodies and potent immune and growth factors. It supports normal cell growth, tissue repair and system-wide rejuvenation that occurs with loss of immune and growth factors.

Most pathogens enter the body through the mucous membranes in the bowel, then pass through to the intestinal tract. The antibodies and immunoglobulins in Colostrum are not absorbed, therefore they remain in the intestinal tract where they attack pathogens before they can penetrate the body's defense system.

Colostrum is *not* a Greek theatre...

In order for Colostrum to be "body ready" it needs to be pre-digested through a lactic acid fermentation that enhances it's bio-availability without damaging its immunoglobulins. Through this specialized process of fermentation, the lactose content is virtually eliminated making it acceptable to most individuals that are lactose intolerant. Additionally, fermentation produces many digestive enzymes that aid in the assimilation of food nutrients. According to William Campbell Douglass, M.D., the protein contained in cultured (lacto-fermented) dairy products are the highest quality available for human consumption.

What's the job description of growth factors?

Medical research has shown that growth factors in Colostrum assist the body to:

- Regenerate normal growth of muscle, bone, cartilage, skin, collagen and nerve tissue
- Burn fat for fuel instead of muscle tissue during dieting
- Build and retain lean muscle
- Synthesize DNA and RNA
- Balance and regulate blood sugar levels
- Heal burns, cuts, abrasions and mouth sores with topical application
- Regulate blood glucose levels and "brain chemicals" providing alertness, better concentration and reduced energy from "glucose lows"

AUTHOR'S NOTE: In facilitating wellness for thousands of clients around the world, one thing that is always a common denominator is the need for boosting the immune system. In doing so however, I have had to take into consideration the compromised digestive and elimination system of clients who are as ill as I was. I attempted several brands of bovine (cow's)

Colostrum, each time discontinuing supplementation when excessive gas, bloating and lower G.I. discomfort occurred. When I started a routine of goat's milk colostrum, it was easily tolerated for myself and most of my clients, especially those with leaky gut, irritable bowel, Crohns, candida, prescription drug withdrawal and multiple allergic response syndrome (MARS). The product I use and recommend is Goatein IG™.

Why is Goatein IG™ superior to other colostrums?

Because Goatein IG™ is the ONLY product on the market containing goat's milk colostrum that is bioactive serum and:

- Contains 100% pre-digested protein
- Contains immunoglobulins (IgA, IgD, IgE, IgG, and IgM), peptides and growth factors
- Their goats are NOT fed pesticides, herbicides, growth hormones or antibiotics
- It can be taken by most people who cannot tolerate cow's milk products because the lacto-fermentation virtually eliminates the lactose content
- Contains all the essential amino acids for optimal health
- Is minimally processed to preserve its original composition, without the use of acid or excessive heat. This allows amino acids, enzymes and beneficial bacteria to remain in their natural form.
- Goatein IG™ is processed at low temperatures and dried on site at the farm to maintain high biological activity. Most other colostrums are produced on farms and then shipped hundreds or thousands of miles for processing.
- Contains 500 mg. Of Goat-Milk Colostrum per serving

- Is high in immunoproteins and contains biologically active Cystine, Glycine and Glutamic Acid in tri-peptide form. These three amino acids produce Glutathione. Glutathione functions as a principal antioxidant, scavenging free-radicals and environmental toxins such as lipid peroxides that can damage and destroy healthy cells.
- It contains important IgF-1 constituents. This form of ImmunoGlobulin has been studied for its ability to stimulate healing, promote muscle growth, and aid in the production of immune cells.
- It contains the powerful, naturally occurring enzymes of protease, amylase, lipase and lactase that contribute to the digestive health and absorption
- Contains probiotics in a base of lacto-fermented goat-milk protein combined with goat-milk colostrum
- Aids in energy production and fat-burning through its lactic acid fermentation process
- Creates proper pH balance in the gastrointestinal tract and other tissues because of the same lactic acid fermentation process

Colostrum: a natural anti-Inflammatory

The following will describe how this natural substance can replace steroids, such as Prednisone, which are used to block inflammation, which can have serious side-effects including bone loss and immune dysfunction. Doesn't it seem ironic that a weakened immune system is one major reason inflammation occurs in the first place, and the drugs used to block it cause the breakdown of the immune system? Colostrum is a safe, effective alternative to steroids, or in conjunction with steroids in acute cases, to block inflammation.

According to D.K. Murphey, a pediatric resident, colostrum forces the differentiation of certain cells so that rapid onset, severe, and extended inflammation do not occur. The initial reaction will probably still occur, but at a greatly reduced rate. He participated in an animal study that demonstrated colostrum's effects on an acute inflammatory process in vivo. Inflammation was challenged with normal saline, pooled aqueous human colostrum, or carrageenin plus colostrum concurrently.

The injection of colostrum demonstrated suppression of a provoked acute inflammatory response. His data directly demonstrated, for the first time, that human colostrum has a biologically significant effect on the inflammatory process.

FYI...

The recommended dose of Goatein IG™ for a healthy individual is 5 caplets per day. I found that for those with compromised digestive systems, and multiple allergic response syndrome (MARS) it first test for compatibility. If the lower dose is well-tolerated then slowly move-up the amount until eventually the maximum recommended dose is tolerated.

For those individuals without a compromised digestive system who choose to enhance their immune system as a pro-active approach, 5 caplets once a day is adequate. For those under excessive stress, athletes, weight-lifters or those regularly exposed to toxic substances, it is best to consume 5 caplets at least twice a day.

Colostrum works best if taken on an empty stomach; at least 45 minutes before or after a meal.

Take Colostrum with room-temperature or chilled water, not with warm water; heat can destroy not only the colostrum but the probiotic content as well.

Colostrum does not require refrigeration but should be kept in a cool, dark and dry environment.

For the frail or elderly, I find it best to suggest 2 caplets three times a day.

For children who are not able to swallow the caplets, it is best to dissolve them in room-temperature water and then mix with a juice of their choice, preferably not a citrus or highly acidic juice.

Colostrum is *not* a Greek theatre...

Colostrum has been found to suppress inflammation through various factors present which offer anti-inflammatory protection.

Colostrum enables rapid elimination or neutralization of microbial pathogens, thus one explanation for it's anti-inflammatory abilities.

According to A. Goldman, as published in the *Annuals of the New York Academy of Science*, the following anti-inflammatory properties are present in Colostrum:

- Agents that double as direct protective agents
- Antioxidants
- Enzymes that degrade inflammatory mediators
- Anti-enzymes
- Cytoprotective agents
- Modulators of leukocyte activations

Autoimmune diseases: colostrum to the rescue!

There exists an excess of 30 different autoimmune diseases and many chronic disorders that are also considered in this category.

Any time the immune system's weaponry turns against itself, stages a battlefront and issues the order to destruct, autoimmune diseases emerge as the casualty of friendly fire.

The statistics for autoimmune diseases in North America alone are staggering; five percent of the adult population is afflicted with one or more autoimmune diseases–two-thirds of those are women. For more specific information about these diseases, refer to my book "Invisible Illnesses" (Freedom Press, 2002).

"Did you know?..."
Many of the healthiest and longest living people in the world have consumed lacto-fermented goat-milk products and attribute their longevity to this remarkable food.

Which autoimmune diseases can benefit from colostrum supplementation?

- **MULTIPLE SCLEROSIS**–a disease in which the myelin sheath that protects the nerves in the central nervous system is destroyed.
- **RHEUMATOID ARTHRITIS AND JUVENILE RHEUMATOID ARTHRITIS**–a disease where the peripheral joints are destroyed.
- **FIBROMYALGIA/MYALGIC ENCEPHALOMYELITIS/CHRONIC FATIGUE**– syndromes that manifest as pain and inflammation of the soft and connective tissue, un-restorative sleep/insomnia, digestive disturbances, excessive yeast and generally parasites, mood swings, fibro-fog, chest pain, swollen lymph nodes, and lack of stamina.
- **TYPE I DIABETES MELLITUS**–a condition in which the cells that produce insulin are destroyed.
- **MYASTHENIA GRAVIS**–receptor cells for the acetylcholine neurotransmitter are destroyed.
- **GRAVES' DISEASE**–a result of the destruction of thyroid stimulating immunoglobulins or anti-thyroid stimulating hormone receptors.
- **ALLERGIES, ASTHMA, MULTIPLE ALLERGIC RESPONSE SYNDROME—**a result of the immune system over-reacting to stimuli.
- **ANEMIA (PERNICIOUS)**–antibodies to gastric parietal cells and vitamin B_{12}; (Hemolytic) anti-red blood cell antibodies.

"You need to know..." Research has shown that regular consumption of cultured (lacto-fermented) dairy products lowers cholesterol and protects against bone loss.

- **LUPUS ERYTHEMATOSUS**–DNA, blood vessels, skin and kidneys are destroyed.
- **AIDS**–T-helper cells are destroyed.

Other autoimmune diseases include, but are not limited to:

- **ADDISON'S DISEASE**
- **CROHN'S DISEASE**
- **CYSTIC FIBROSIS**
- **INSULIN-RESISTANT DIABETES**
- **RHEUMATIC FEVER**
- **ENDOMETRIOSIS**
- **HASHIMOTOS'S THYROIDITIS**
- **PSORIASIS**
- **CELIAC DISEASE**
- **ANKYLOSING SPONDYLITIS**

Part six

Understanding the digestive operating systems

THE CORE OF THE digestive system is the gastrointestinal tract (GI). The GI tract is a hollow tube that begins at the mouth, where food enters the digestive system, and ends at the anus, where fecal waste products are expelled. Between the mouth and anus are the esophagus, stomach, small intestine and large intestine. The organs of the liver, pancreas and gallbladder support this diverse and essential tract. Collectively, these organs make up the digestive system.

What makes up the gastrointestinal digestive system?

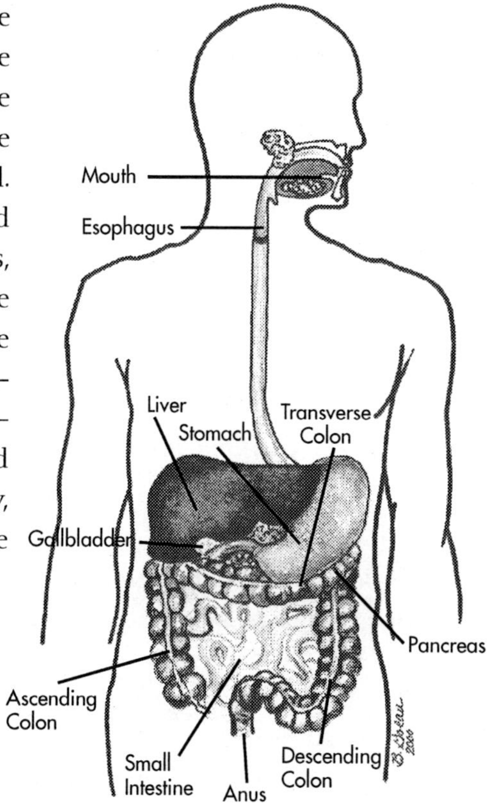

The three stages of the digestive process

STAGE ONE: Food is chewed, therefore creating larger surfaces for enzymes and enzyme-containing saliva to begin their work.

STAGE TWO: Once swallowed, the food continues its journey into the pharynx and esophagus, which do not contribute to the digestive process. However, they provide the route for the food to reach the stomach.

STAGE THREE: When food arrives in the stomach, it will then trigger the production of hydrochloric acid. This process will take approx. 45 minutes in young healthy adults. However, when people reach the age 40 and beyond, the hydrochloric acid production is slowed significantly, and after age 60 it is minimal if at all. This accounts for the difficulty many seniors have in digesting proteins.

Introduction to enzymes by Beata Golau, enzyme specialist

Enzymes are Life itself. All living things are dependent on enzymes for their biochemical and physiological functions. Vitamins and minerals are the building blocks of the body, however, enzymes are the ones who have the capacity to do the actual work of delivering nutrients and acting as the body's internal operating system. Lack of enzymes can be likened to having all the building materials (vitamins and minerals) sitting on an empty lot without any workers placing them where they need to be to become a structure.

Since the early 1900's, enzymes have been systematically removed from our food supply in order to extend the shelf-life of food.

Within each food are the enzymes to digest that particular

food. As an example, when you drop an apple, a bruise forms where it hit the ground. While we call that bruise spoilage, it is actually digestion occurring due to the activation of the naturally occurring enzymes in the apple. Food producers must remove those same enzymes from our food supply in order to achieve an extended shelf-life. This translates into removing the actual life-giving force in our food in order to keep it for longer periods of time without it "spoiling". Therein lies the problem, and the necessity for supplemental enzymes. Your health is the price being paid for that longer shelf-life.

While vitamin and mineral deficiencies produce more acute symptoms, food enzyme deficiencies surface as more chronic degenerative conditions. This is very evident when statistically it is estimated that over 62 million North Americans alone suffer some form of a digestive disorder.

Our bodies were not intended to bear the entire burden of digesting our food. What enzymes we don't get from our food, our body has to replace from another source. This ends up being a case of "robbing Peter to pay Paul," often resulting in undigested food. If our body has to make a choice between digesting food or running an organ, it will always run the organ to preserve life–now the undigested food becomes a toxin! Immune responses come into play to try to filter the toxin out of the body. Now, suppose a cancer cell needs to be defused. Will the body be able to do that if it is already busy detoxifying undigested food? If you consider that this goes on, and on a daily basis in an enzyme deficient diet, it is a tremendous

"Food for thought..."

What you crave is most likely what you are NOT digesting. The hypothalamus (part of your brain) signals your stomach that you need a specific nutrient, however, if you are not digesting that nutrient, your hypothalamus will once again send out a request... so... there you go again...wanting more of the same thing, still not digesting and still needing that specific nutrient... AND the beat goes on.

amount of extra work the body must perform. At some point, the extra burdens placed on the body take their toll, thus the advent of auto-immune diseases and chronic disorders.

Fortunately, we have nutritionally aware medical doctors and health care professionals who have recognized this situation and developed enzymatic therapies that deliver nutrients past an incompetent digestive system. There are also now various tests that are able to determine if the body is breaking-down various food groups (carbs, fats and/or proteins) and if those nutrients are being assimilated and delivered to the appropriate places to carry out their jobs. Thus, instead of "hushing-up" symptoms with an antagonistic chemical, we are often able to gradually restore normal function...the goal of enzymatic therapy.

Enzymes: there is NO life without them!

I'm including information regarding the vital importance of enzymes because I would be derelict in my teaching as a doctor without mentioning their role. Many individuals spend a great deal of money, energy and time buying the best food and supplements; yet do not gain the benefit from the products because they lack digestive enzymes.

Every day over 200,000 people, and that's just Americans, stay home from work because of some form of digestive disorder. It is estimated that digestive disorders/diseases are responsible for $50 billion in lost work and wages, uncollected taxes and health care payments. Statistics indicate that one in every two Americans experiences digestive disorders such as indigestion,

FYI...
It is important to note that although digestive enzymes play a vital role in the digestive process, they also directly assist the defense mechanisms of the immune system.

ulcers, bloating, heartburn/acid reflux, gas, irritable bowel syndrome, and leaky gut syndrome.

Primarily enzymes are a catalyst; they're energized protein-like substances that accelerate and precipitate the hundreds of thousands of reactions in the body that control life's processes. Digestive enzymes break-down food particles for storage in the liver. One physician calls enzymes the "Sparks of Life."

According to Lita Lee, Ph.D., an enzyme specialist, "Depletions of plant enzymes lead to a host of chronic diseases that could in part be avoided if we provided the body with the enzymes it needs. As it is, we are not aware of enzyme deficiencies because they take so long to manifest. When there are signs, the body is already in a state of exhaustion." If you're the victim of an invisible illness, as described in my book "Invisible Illnesses" (Freedom Press, 2002), you surely have a compromised digestive system to some degree. Dr. Lee believes, "If you have fatigue, allergies, bloating, gas, constipation, indigestion, or any symptom of undigested food, then you most likely have an enzyme deficiency."

Consequences of enzyme depletion

Deficiencies of the main digestive enzymes, protease, amylase, lipase, cellulase and disaccharidase (sucrase, maltase, lactase) include the following signs and symptoms:

- **PROTEASE DEFICIENCY (digests protein)**–anxiety, cold hands and feet, edema (water retention) frequent infections, colon disorders, kidney disease, osteoporosis, bone spurs, blood clots, hypoglycemia (low blood sugar)
- **AMYLASE DEFICIENCY (digests carbohydrates)**–muscle soreness and pain, joint stiffness, thick blood, psoriasis, hives, rashes, eczema, herpes

- **LIPASE DEFICIENCY (digests fats)**–obesity, difficulty losing weight, diabetes, dizziness, high blood pressure, cardio-vascular problems, gallbladder dysfunction
- **CELLULASE DEFICIENCY (digests fiber)**–acute food allergies, candida, leaky gut syndrome, facial pain/twitching/paralysis, gas, bloating
- **DISACCHARIDASE DEFICIENCY (digests sugars)**–attention deficit disorder (ADD), environmental illness/multiple allergic response syndrome, asthma, bronchitis, chronic diarrhea, dizziness, hyperactivity, insomnia, mood swings, fibro-fog (brain-fog), seizures

Enzymes: their names and what they do

I looked, used and tested many forms of digestive enzymes and found that what I was really looking for was a combination that contains a full-spectrum, is 100% vegetarian, and included the following:

- **PROTEASE BLEND:** Three powerful protein-digesting enzymes, including peptidase, which is capable of breaking down proteins at different optimum pH levels along the digest tract. A protease that was capable of breaking down proteins found in meat, cheese, nuts, and eggs. A protease that assisted in the reduction of infection and inflammation.
- **AMYLASE:** To break-down and digest carbohydrates, starches, and sugars contained in potatoes, fruits, vegetables, and many snack foods–important because it works in harmony with other digestive enzymes to prevent the proliferation of dead leukocytes (white blood cells).

- **GLUCOAMYLASE:** Because this enzyme further breaks down and digests carbohydrates, starches, and sugars; in particular, maltose sugar and polysaccharides.
- **LIPASE:** Because of it's powerful fat digesting properties. Lipase breaks down fats such as those in meats, dairy, oils, nuts, and seeds–important for weight management and circulatory function.
- **LACTASE BLEND:** A blend so powerful that it contains a full spectrum of enzymes specifically designed to break-down and digest dairy products as well as the accompanying factors of lactose (milk sugar), and protein. A blend that would ease or completely eliminate "lactose or dairy intolerance" and the accompanying symptoms of gas, bloating, cramping, indigestion, and diarrhea. A blend that would be at least three times stronger than the dairy-digesting enzymes found in the stomach.
- **BROMELAIN:** A unique enzyme extracted from the fruit and leaves of the pineapple plant. Unique because it contains several proteolytic (systemic) enzymes which work well in different digestive environments. The proteolytic composition of this enzyme has been used for centuries around the world by health care professionals because of its additional ability to break-down protein and reduce inflammation–a natural anti-inflammatory.
- **PAPAIN:** Used historically to control/relieve chronic diarrhea, it helps to break-down proteins in the body but additionally through other components degrades fats and carbohydrates. It also has the ability to treat gastrointestinal

"You need to know..."
Enzymes are more heat-sensitive than vitamins and are destroyed in the cooking process at temperatures above 118°F. They are also destroyed by microwaving, pasteurizing or canning. Therefore, in this day and age of fast and convenient food, it becomes even more important that we eat some raw foods daily and take enzymes.

discomfort due to intestinal parasites, fights allergies and infections, used in many types of surgery to decrease inflammation, pain and swelling and actually has a sedative effect.

- **ALPHA GALACTOSIDASE:** Because of it's ability to digest complex sugars in whole grains, beans, legumes, cabbage, broccoli, and cauliflower (cruciferous vegetables). The fibrous nature of these foods poses digestive challenges for many people because humans do not produce this enzyme that is required to break them down. It's also exceptional in eliminating bloating, cramping, and intestinal fermentation by the breakdown that occurs before they reach the lower intestine.
- **CELLULASE:** A complex of three different enzymes that converts fiber cellulose (one of the basic components of the cell wall) to glucose. It helps to break-down the soluble parts of fiber contained in raw fruits, vegetables, and whole grains. Cellulase is not made in the body and can only be obtained from our food or supplementation.
- **ISOLASE™:** An enzyme blend designed to increase absorption of isoflavones (most commonly found in soy). Clinical trials show that this enzyme blend can actually double the isoflavones in humans. It has also shown cancer-preventive properties as well as positive effects on osteoporosis, and cardiovascular disease.
- **MALT DIASTASE:** This enzyme is also known as maltase, and helps to break-down maltose, malt, and sugars found in grains.
- **CERECALASE™:** A proprietary blend of enzymes designed to break-down carbohydrates and fibers found in many fruits, vegetables, and grains. CereCalase™ contains Hemicellulase (which breaks-down hemi-cellulose carbohydrates). Phytase (which breaks-down carbohydrate phy-

tates such as those in soy, gluten, wheat, oats, rye, and barley), and BetaGlucanase (used to digest polysaccharides and fibers). CereCalase™ helps to break-down all non-starch polysaccharides (NSPs), which are common in many vegetable foods including grains, beans, and herbs.

- **SOD/GLIACIN (VEGETABLE SOURCE):** Superoxide Dismutase (SOD) is a powerful enzyme that revitalizes cells and reduces the rate of cell destruction. It helps to neutralize free-radicals in the body and reduce inflammation. SOD also aids in the assimilation of zinc, copper, and manganese, and is currently being explored as an anti-aging treatment. It is contained in foods such as barley grass, broccoli, Brussels sprouts, cabbage, and wheat grass.
- **L GLUTAMINE:** Is important in maintaining a healthy digestive tract. It is used therapeutically for the treatment of peptic ulcer and intestinal mucosal deterioration (leaky gut syndrome).

"You need to know..."

I take one capsule of Caprazyme™ at the beginning of my meal and one after the meal; this gives me the best enzymatic action and acts as a sort of "time-release." For children and pets, the capsule can be opened and sprinkled on the food; it's odorless, taste-less and easily disguised. I have a 4 lb. toy poodle (Ms. Scarlette) with digestive problems; without sprinkling enzymes on her food, Ms. Scarlett would belch and burp after every meal. I've suggested this to many of my clients and they report amazing results in the health of their pets. Dosage for animals varies according to weight, consult your health care professional for exact dosage.

I also wanted a digestive enzyme formula that included a bioactive botanical blend, as most digestive enzyme formulas do not. It needed to contain the following:

- **GINGER ROOT:** Because it's been used for over 5,000 years for it's ability to fight inflammation, soothe the stomach, cleanse the colon, reduce spasms and cramps, and stimulate general circulation. It also contains generous amounts of Vitamin A as well as numerous phytochemicals such as beta-carotene, camphor, citral, curcumin, and capsaicin. It is a powerful protein digesting enzyme.

- **PEPPERMINT (WHOLE LEAF):** Because of it's ability to increase stomach acidity that aids the digestive process. Also because it has an anesthetizing affect on the mucous membranes and the GI tract, thus assists in relieving intestinal pain. It is used world-wide in the treatment of colic, diarrhea, indigestion, IBS, nausea, and poor appetite. Peppermint also has an antispasmodic effect on the bowel walls, causing them to relax.
- **ALOE VERA (WHOLE LEAF):** Because of it's unique capacity to soothe mucous membranes, act as an astringent, anti-fungal, anti-bacterial, and anti-viral. Aloe Vera has been shown in studies to lower cholesterol, reduce inflammation, soothe stomach discomfort (including ulcers and acid reflux), and act as a natural stool softener.
- **NARINGIN:** Is a botanical rarely used and yet has the unique ability to act as a powerful up-take mechanism to enhance nutrient absorption. It is derived from grapefruit and is a flavonoid that contains numerous anti-oxidants and carotenoids.

FYI...
Cancer is now believed to be the result of decreasing tolerance to toxic substances, poor elimination and depleted nutrition, and...enzymes deficiencies play a major role.

Finally, a product was developed that not only met, but exceeded, all my criteria, CapraZyme™. It's nice to know someone does listen to health care professionals.

Part seven

Cultivating a garden of health—in the gut that is!

THE FOLLOWING ARE TERMS that are imperative in the understanding of the relationship of the gut and the factors for healthy intestinal ecology.

Malabsorption

A common symptom of a compromised digestive system, specifically the gut is not unlike a garden or potted plant that is so root-bound it can no longer absorb nutrients to promote new, healthy growth. The more beneficial bacteria present in our intestinal garden to fertilize the beneficial bacteria, the more efficiently the body can absorb minerals and nutrients from the food we consume.

Probiotics

"Friendly" bacteria that exists in our intestinal tract. These beneficial bacteria re-inoculate the intestinal ecology after taking substances such as antibiotics that kill most bacteria (the good, bad and the ugly). Another example of good bacteria being destroyed is when you encounter a bout with diarrhea. Probiotics are bacteria similar to the "good" bacteria in

"Did you know?..."

In studies performed on hospitalized seniors who were victims of constipation, over 95 percent reported alleviation of symptoms after taking a high-potency probiotic. In another study at Augsburg College in Minneapolis, probiotic capsules were given to people prior to traveling to Mexico, Nepal, and Guatamala. The subjects reported no incidents of diarrhea or intestinal disorders, whereas, their partners who did not receive the probiotic had intestinal disorders despite normal precautions while traveling to a third world country.

our intestines. Once taken, the "live" probiotic bacteria formula (like acidophilus and bifidobacterium) must survive stomach acids before they reach the intestines. When you add a probiotic, it's much like re-seeding your lawn or garden when it becomes sparse; it simply means you are actually introducing new seeds (bacteria) to assist in increasing the existing population of good bacteria.

Intestinal flora— the body's landscape

There are over 400 different kinds of bacteria living in our Gastro-Intestinal tract (GI). The human GI tract is supposed to contain 85% good bacteria (friendly) and 15% bad bacteria "unfriendly". Unfortunately, most people in the western world have the opposite.

The human body is designed to serve as home to trillions of good bacteria. These tiny friends work hard, helping our immune system ward-off viruses, bad bacteria and fungi. They also play a role in creating "B" Vitamins and certain enzymes. In order to do this, probiotic populations must be high, the strains of microorganisms viable, and of top quality. They must be specifically cultured to survive stomach acids and the entire digestive process. They must be able to form colonies in the intestinal tract where they aid the body to absorb nutrients and establish the proper pH. They need to be able to implant in most colonic environments, but especially thrive in acidic conditions.

Earlier in this book I dealt with the importance and superior quality of goat-milk products and why they are so beneficial to humans. However, until now, there were no probiotics being made by placing these organisms into a nutrient-rich bed of fresh organic goat's milk. And...since goat's milk

"You should also know..."

- According to nutritionist Robert Gray, "A less-known but equally important function of the friendly bacteria within the intestinal tract is to provide important nutrients for building the blood. Many cases of "tired blood" would benefit by taking steps to insure a continual high concentration of beneficial bacteria within the intestinal tract than by taking iron tonics. A healthy intestinal flora will produce several times as much of many B vitamins as are present in a well-balance whole-food-diet."

- Probiotics, such as acidophilus, actually adhere to the intestinal walls and vagina and prevent disease-causing bacteria from taking hold. When these "good bacteria" take-up residence in the intestines and leave "no vacancy", we're protected. However, when there's a "vacancy" (lack of good bacteria) the invading bacteria can come in and quickly occupy the premises—thus the body is left without full protection.

- The following can contribute to a condition of intestinal and vaginal flora imbalance (the friendly bacteria being out-numbered), causing substances to not be excreted from the body:

  Overuse of antibiotics
  Cortisone
  Prednisone
  Oral Contraceptives
  Excessive Stress
  Antihistamines
  Aspirin
  Excess Sugar Consumption
  Coffee

- There are a total of one-hundred trillion bacteria that co-exist in our digestive system; some harmonious, some antagonists. Their total weight accounts for approximately 4 pounds; the size of an average liver. These bacteria raise or lower our risk of disease by either strengthening or weakening our over-all immune system. They live and thrive in your intestinal garden at about 98.6 degrees F and require a warm, dark, moist environment...a perfect place to "bloom"—let's be sure we're planting "friendly" bacteria.

- "Friendly" bacteria, probiotics, help increase our resistance to food poisoning as well. In 1993, The United States reported 20 to 40 million cases of food poisoning, although the FDA estimates the accurate total to be closer to 80 million.

is the most widely consumed and tolerated milk in the world, and is credited with more health benefits than almost any food, doesn't it make sense that someone would make a probiotic from goats milk? Even the Journal of American Medicine said, "Goat's milk is the most complete food known." Just image, taking this uniquely complete food which contains vitamins, minerals, electrolytes, trace elements, enzymes, protein, fatty acids, easily utilized by our bodies with ease, and....making the first probiotic from pure goat's milk; CaproBiotics™ did just that. This unique formula contains over 24 billion beneficial microorganisms per serving and is cultured in fresh, pure goats milk. Their goats are never fed any pesticides, herbicides, antibiotics or growth hormones. It contains naturally occurring "live enzymes" and specially cultured probiotic strains make for added stability and viability.

The six multiple strains of probiotics contained in CaproBiotics™ include:

- **LACTOBACILLUS ACIDOPHILUS:** The most well known lactic acid bacteria, and is a necessary inhabitant of the intestinal tract. CaproBiotics™ *contains two strains* of these bacteria with at least 12 billion CFU (Colony Forming Units) per serving. The beneficial properties of L. acidophilus include: production of enzymes such as Protease (digests proteins), Lipase (digests fats) and Lactase (digests lactose). It produces "B" vitamins, a natural antibiotic, and Acidophilus inhibits the growth of pathogenic bacteria.
- **LACTOBACILLUS CASEI:** The most potent of all lactobacillus strains. There are at least 12 billion CFU (Colony Forming Units) per serving of this microorganism in CaproBiotics™. It's also a natural producer of lactic acid that aids in the

digestion of a wide range of carbohydrates and other nutrients.

- **BIFIDOBACTIRUM LONGUM:** A potent cleanser of the body and strengthener of the immune system. It is credited in helping to reduce cholesterol levels.
- **LACTOBACILLUS BULGARICUS:** A natural producer of lactic acid that also contains strong natural antibiotic properties thus is a strong immune system enhancer.
- **STREPTOCOCCUS THERMOPHILUS:** This is the strain often used in yogurt and Kefir. It possesses anti-carcinogenic and anti-tumor properties and helps the digestion of milk sugar. This also contains natural antibiotic properties.

"Food for thought..."

The next time you go on a family outing that includes; camping, picnicking or outdoor gatherings on warm days, take a high quality probiotic like Caprobiotics™ as added insurance against food-poisoning. Taking a probiotic doesn't give us permission to be lax in proper preparation and refrigeration of our food, what it does provide is added protection. Probiotics provide protection by not being a good host to the invading microbes. Contrary to common belief, probiotics do not kill invading microbes, they change the intestinal landscape by secreting large amounts of acids (acetic, formic, and lactic acids), which then make the terrain unsuitable for pathogens and they "do themselves in."

Probiotic formulas are not only for illness; they present us with the perfect supplement to maintain health – the essence of probiotic means "for life or helpful for life"...how does your garden grow?

Conclusion

IT IS MY HEARTFELT PRAYER that this book will be an important catalyst for a greater understanding of the magic of goat-milk products and their life-giving and life-enhancing God given abilities.

The quality of life of millions of people throughout the world is at stake. Billions of dollars in our dwindling health-care system can be saved if we can effectively provide natural products for health-care, rather than drugs for symptom-care that eventually leads to drug-management.

To the conventional medical experts that kept echoing I was "within normal range" while all along my liver was ready to shut-down from self-poisoning I say: "I am thankful to be alive–living well *is* the best revenge–and I plan to continue working with the same passion, faith, and conviction that allowed me to recover. My goal is to share my experiences, my knowledge, my research and my message throughout the world."

To Your Health, Naturally!

Survey

MY WORK AND RESEARCH CONTINUES into the health benefits of goat milk products. I welcome your input and stories of how goat-milk products facilitated your road to recovery or were used to maintain health.

I am also interested in adaptations in menus and other uses you've found for products mentioned in this book.

Now that you're on your "Whey" to health, my hope is that you'll share and, therefore, have an opportunity to make a positive contribution in the lives of others.

Please send your stories one of two ways:

Mail:
Dr. Gloria Gilbère
Survey
P.O. Box 1565
Sandpoint, ID 83864-1565 USA

Or Email:
doctorgilbere@hotmail.com *(Please use ONLY for surveys)*

Afterword

IT ISN'T EVERY DAY, THANK GOD, that a doctor has the opportunity of turning a personal life-threatening health challenge into an opportunity–one in which to educate and share experiences and research with peers and patients. As a survivor of a life-threatening accident and the subsequent invisible illnesses, I became intent on offering practical non-drug solutions for these often-misdiagnosed syndromes–solutions that allowed me, and those I consult with, to once again achieve wellness and restore quality of life, Naturally.

In my book *Invisible Illnesses* I introduce the physician and the patient to natural therapies and life-style modifications for overcoming the disorders of invisible illnesses. Granted, doctors perform daily miracles by repairing physical damage and removing, and often replacing, diseased organs. However, conventional medicine has little to offer an individual with digestive diseases, environmental illness and chronic disorders. Physicians are trained to "turn-off" a reaction or symptoms with drugs for symptom-care, but in the process, can perpetuate a self-destructive cycle masking the root causes.

I refused to accept the only too-often-heard echo of conventional medicine "everything is within 'normal' range." I ventured out of the "norm" and employed life-style changes and therapies that have withheld the test of time. The products referred to in this book played a large role in my recovery…and that's information too grand to keep quiet.

Today, Thank God, I have resumed my busy private practice, I lecture, teach and consult in over 12 countries and still make time to write and do regular radio and T.V. appearances. Yes, my life has changed forever, it's changed because I have a greater appreciation for what my clients experience, it's changed because the life-style changes that facilitated my recovery are the same ones I continue to employ to sustain wellness, Naturally.

We need to keep in mind that healing is an art as opposed to a science. Here's the dictionary's comparison of science to art: *"A science teaches us to know, and an art to do, and the more perfect sciences lead to the creation of corresponding useful arts. Art is knowledge made efficient by skill."*

I am proud to be in the profession of "The Healing Arts."

Resources

TO ORDER DR. GILBÈRE'S BOOKS, *I was Poisoned by my body*, *Invisible Illnesses*, *Pain/Inflammation MATTERS* and *5 years AFTER I was Poisoned by my body*, call Toll-free (888) 352-8175 or email request to: info@drgloriagilbere.com.

For bulk or trade orders contact the publisher:
Freedom Press 1-800-959-9797 or
E-mail: info@freedompressonline.com

To locate a retailer nearest you for products mentioned in this book contact:
Mt. Capra Products (makers of premium goat milk products and nutritionals, and a retailer of Dr. Gilbère's books)
279 SW 9th Street
Chehalis, WA 98532 USA
1-800-574-1961
www.mtcapra.com

About the author

Gloria Gilbère, N.D., D.A.Hom., Ph.D.

DR. GILBÈRE earned doctorates in both naturopathy and natural health. She's been in healthcare for over 25 years. She is:

- Member–American Academy of Environmental Medicine
- Diplomate–Academy of Homeopathy
- Fellow–The American Naturopathic Medical Association
- Fellow–American Society of Safety Engineers
- Board Member–Fibromyalgia Coalition Int'l
- Medical writer and researcher

Dr. Gilbère consults with clients around the world, both via telephone and at her two health centers in north Idaho. Her services include clinical work, telephone consultations with clients, researchers and physicians. She also consults with neutraceutical companies in the research and development of new products.

She teaches, lectures and consults worldwide, writes regularly for newspapers and trade magazines and is a frequent guest on radio and T.V.

Dr. Gilbère is a keynote presenter and has conducted seminars on varied disciplines of wholistic health, multiple allergic response syndrome, Fibromyalgia/chronic fatigue, colon disorders, leaky gut and chemically induced immune system disorders. Her work has taken her throughout the U.S., Mexico,

Canada, China, Indonesia, Singapore, Argentina, Brazil, Spain, United Kingdom, Switzerland, S. Korea and Greece.

She is internationally respected as a natural medicine researcher, environmental health consultant, author, and an authoritative influence in the discovery of the causes, effects, and natural solutions for leaky gut syndrome, digestive and chemically induced immune system disorders.

For information regarding her consulting services, speaking engagements, interviews or to schedule a consultation by telephone or at her office, Call: (360) 352-3646 (7:30 to 3:00 Pacific Time, M-F) Fax: (208) 265-1777 or Email: info@drgloriagilbere.com (this email address is for inquiries ONLY, please do not ask for health advice). Webpage: www.drgloriagilbere.com

Bill E. Goat

"It's <u>goat</u> to be good...naturally!"